Medical Scientists

Joanne Jalbert

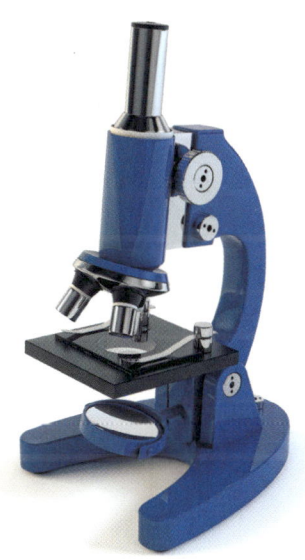

Series Editor **Casey Malarcher**

Level 1 - ❸

Medical Scientists

Joanne Jalbert

Series Editor: Casey Malarcher
Acquisitions Editor: Anne Taylor
Copy Editor: Liana Robinson
Cover/Interior Design: Highline Studio

ISBN: 978-1-943980-35-2

10 9 8 7 6 5 4 3 2 1
22 21 20 19 18

Photo Credits

All photos are © Shutterstock, Inc.
Wikimedia Commons: 23 top

Contents

What Is a Medical Scientist?

Is science your favorite subject in school?

Do you like doing science experiments and helping people?

If so, maybe you should become a medical scientist.

Medical scientists learn about the body and how it works.

A model of the human heart

The layers of ▶
the skin

There are different kinds of
medical scientists.

Most choose one subject to study.

For example, some medical scientists study diseases of
the skin.

◀ A medical scientist
looking through a
microscope

5

Weighing chemicals

Some medical scientists do experiments to find new medicines.

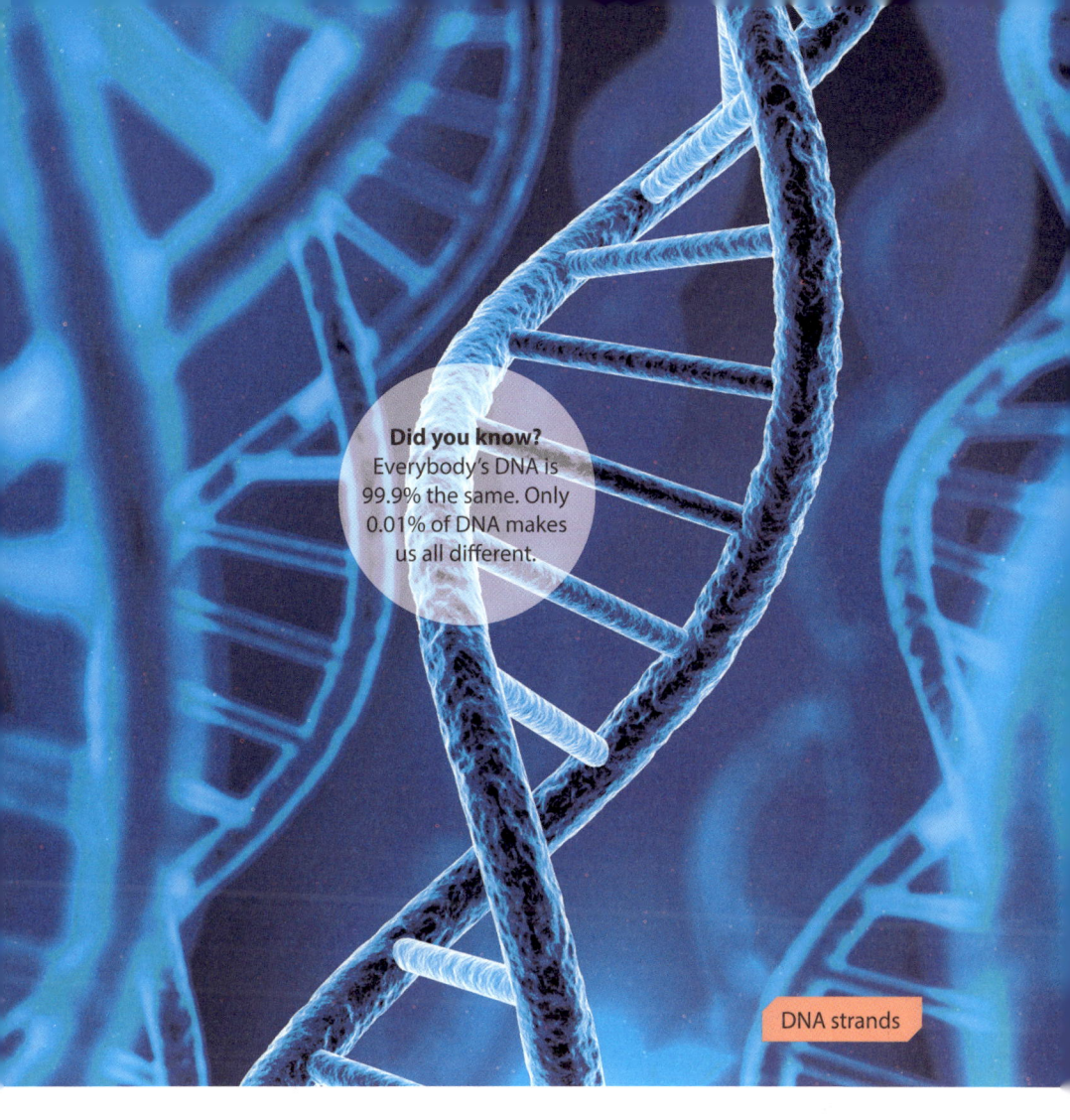

Did you know?
Everybody's DNA is 99.9% the same. Only 0.01% of DNA makes us all different.

DNA strands

Some scientists study parts of the body that are too small to see with our eyes, like DNA.

DNA tells our bodies how to grow.

It is inside every part of our bodies.

Medical scientists study how some people's DNA can make them sick.

In the Laboratory

To become a medical scientist, a student goes to university for many years.

Medical science students study chemistry, biology, and math.

Students also spend many hours in laboratories.

Studying science

A laboratory is a room where people do science experiments.

In laboratories, medical scientists use many things for experiments.

Things that are often used for experiments include a microscope and test tubes.

◄ test tubes

◄ microscope

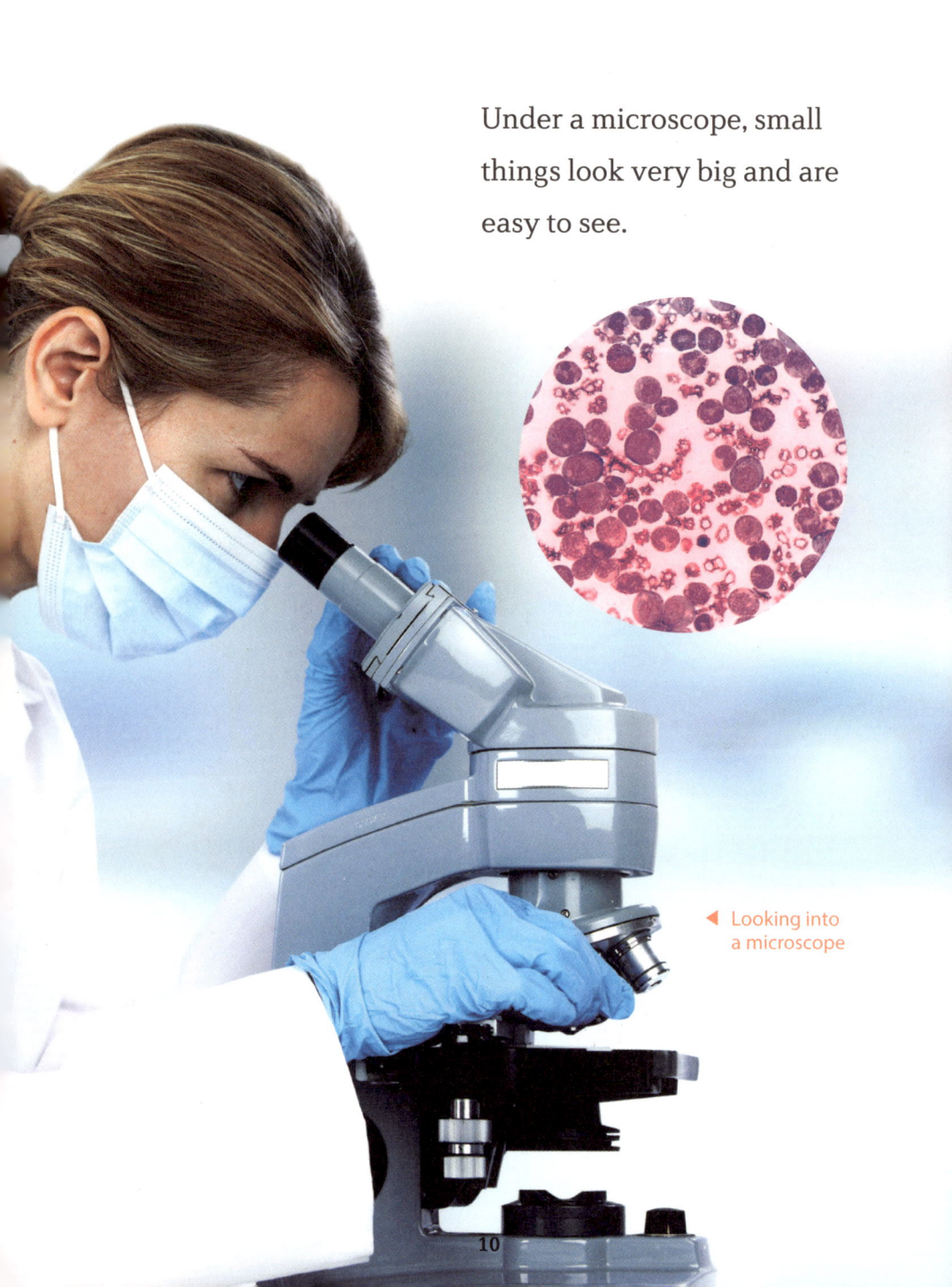

Under a microscope, small things look very big and are easy to see.

◀ Looking into a microscope

Our bodies are made of many cells.

Cells are very small, but we can see them under a microscope.

Some medical scientists study cells to learn about diseases and how the body grows.

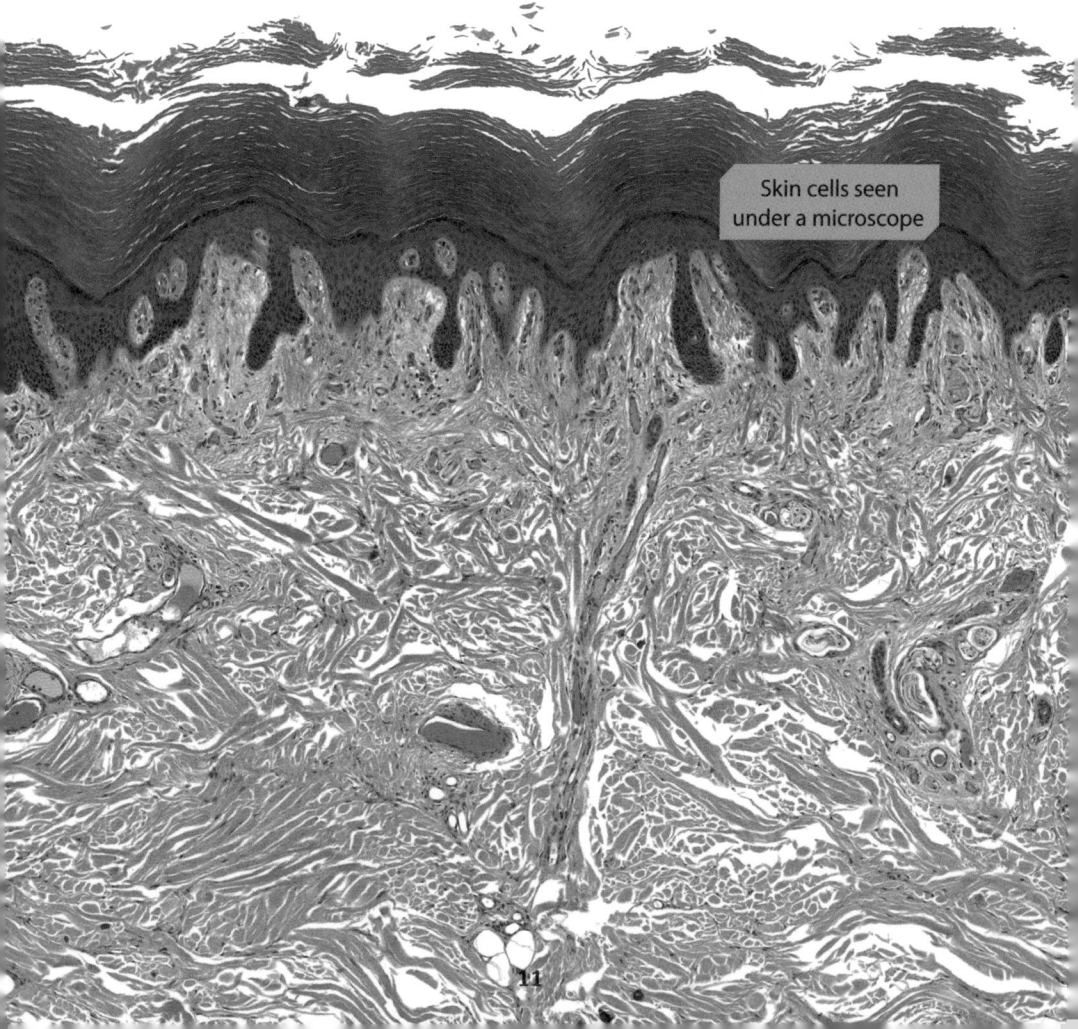

Skin cells seen under a microscope

Scientists seem to find new diseases all the time. In the laboratory, medical scientists make new medicines to fight these new diseases.

Experimenting in a lab

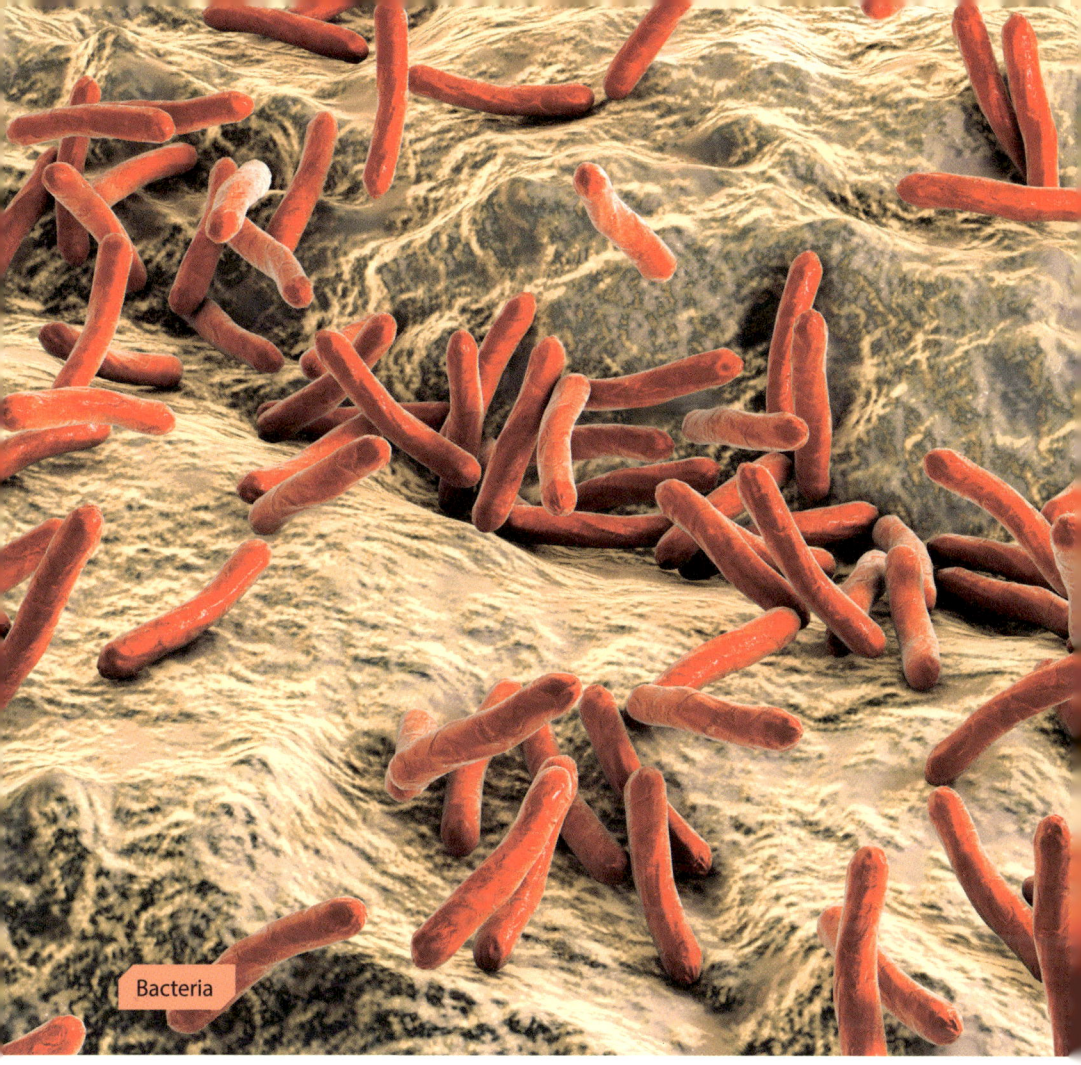

Bacteria

Medical scientists can grow bacteria.

Then they do experiments on the bacteria.

It helps them understand the bacteria.

If they understand bacteria, then they can start to make a medicine to kill the bacteria.

Animals used in medical experiments

Medical scientists test new medicines by using animals.
Scientists do experiments with animals to see if the
medicines are safe.
In the laboratory, scientists try to keep animals safe, too.
They don't want animals to be hurt.

Danger in the Laboratory

In a laboratory, there are many dangerous things.

Gloves keep hands safe.

So medical scientists always wear gloves.

Medical scientists wear special glasses, too.

These keep their eyes safe.

Safety goggles ▶

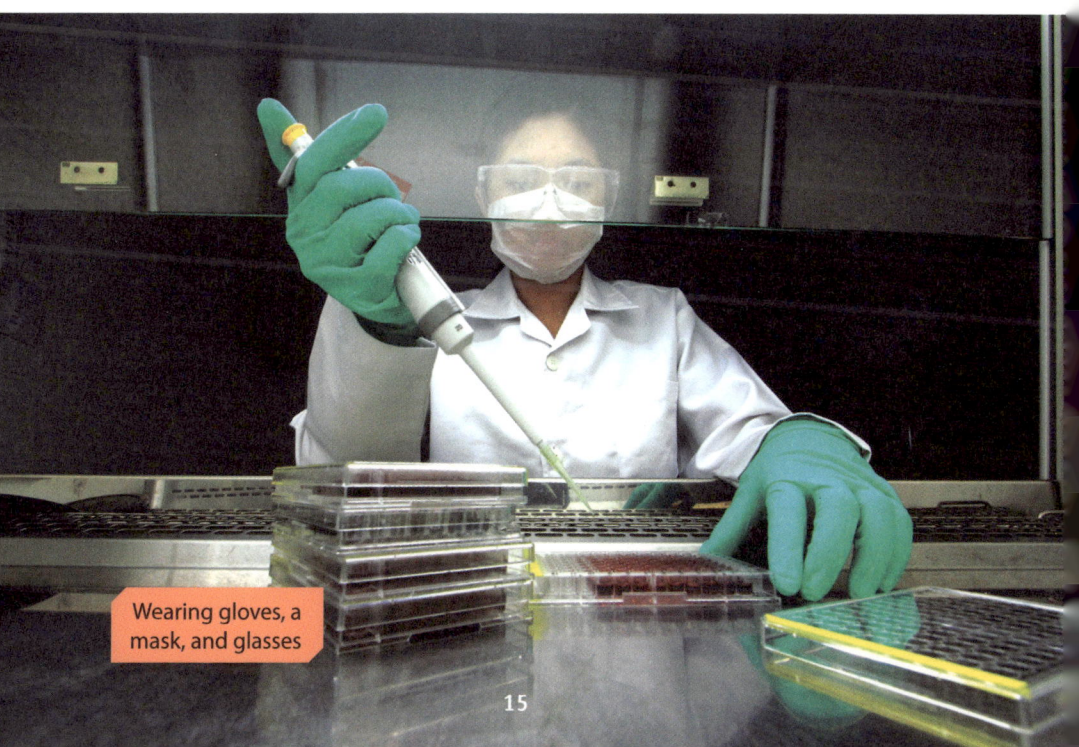

Wearing gloves, a
mask, and glasses

An eye wash in a lab

This machine cleans a person's eyes.

You can see these machines in most laboratories.

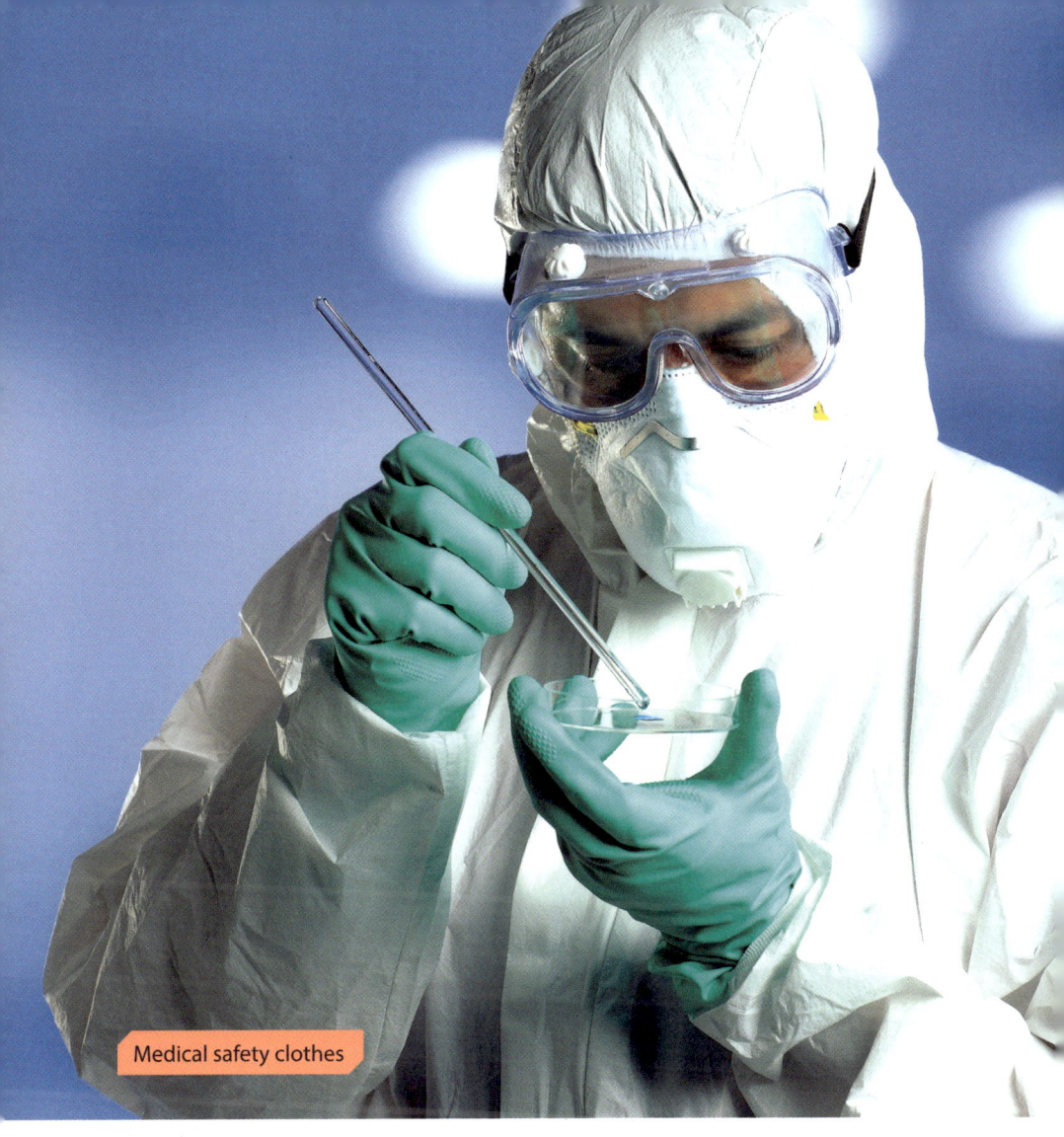

Medical safety clothes

Some medical scientists wear special clothes and masks.
These keep them safe in the laboratory.

Medical Scientists: Helping People

Medical scientists work hard because they like helping people.

If you hurt your hand, medical scientists can use cells to grow new skin for your hand.

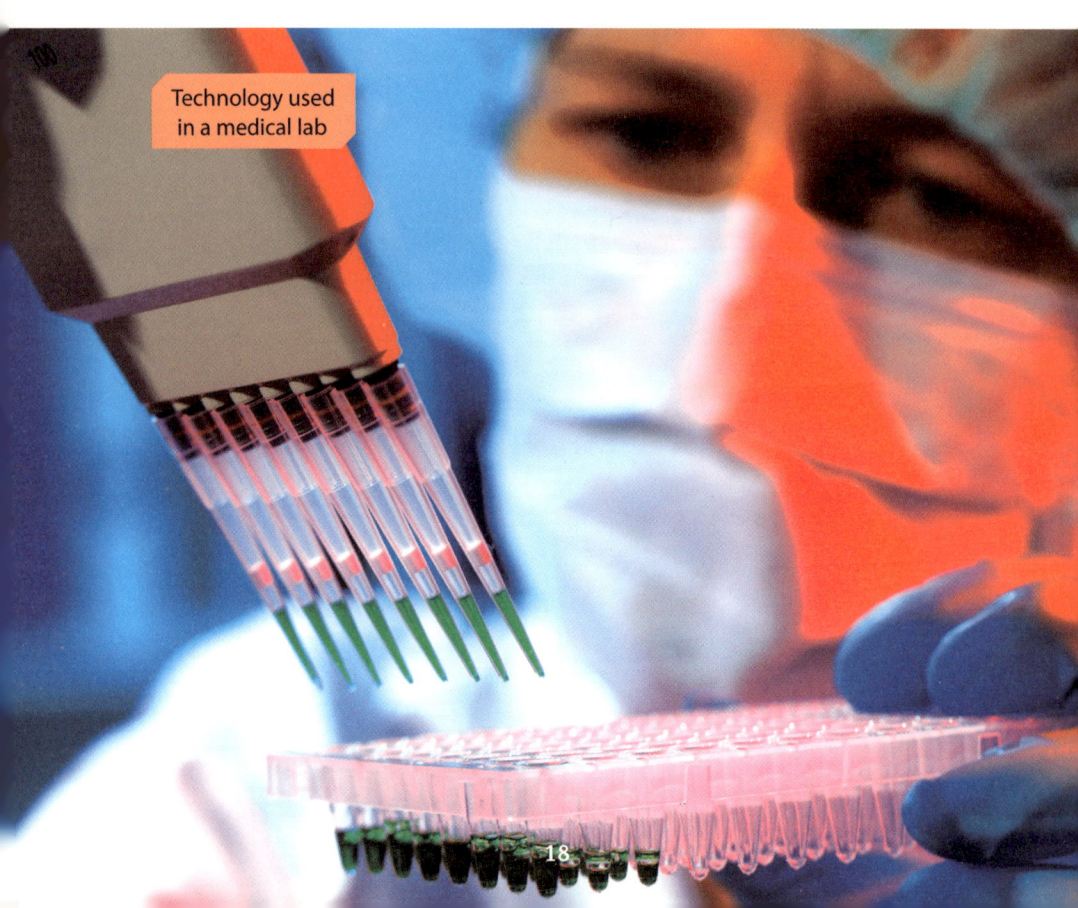

Technology used in a medical lab

Thanks to medical scientists, we can put little machines into people's bodies when a person's heart doesn't work.

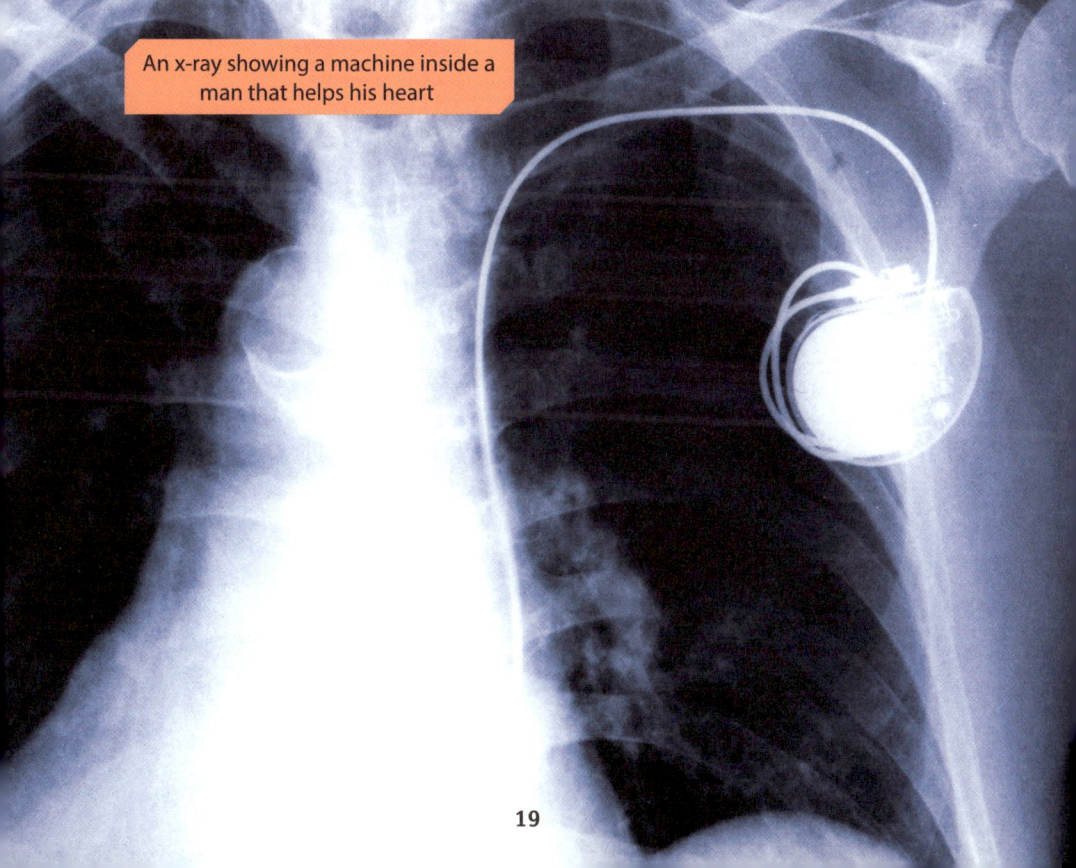

An x-ray showing a machine inside a man that helps his heart

Sometimes, people lose their arms or legs.

Medical scientists make arms and legs that help these people.

The new arms and legs they make can do many things.

A prosthetic hand

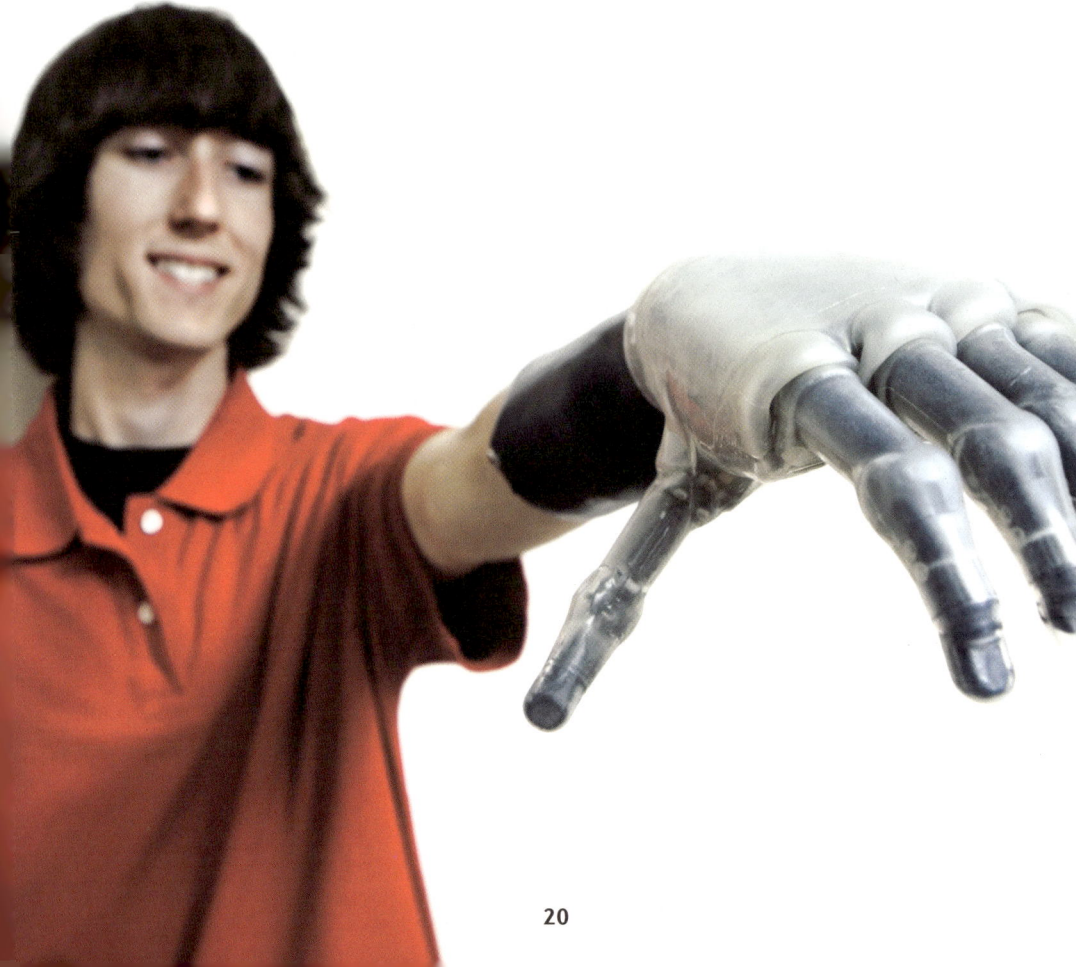

A medical scientist studying a sample in a petri dish

We can thank medical scientists for many things that help people.

A Bright Future in Medical Science

The future of medical science is exciting.

Using computers in medical research

◄ Dr. Craig Venter

Will scientists grow ► babies in labs?

Dr. Craig Venter is an American medical scientist.

He says he can now make cells.

In the future, do you think medical scientists will "make" people?

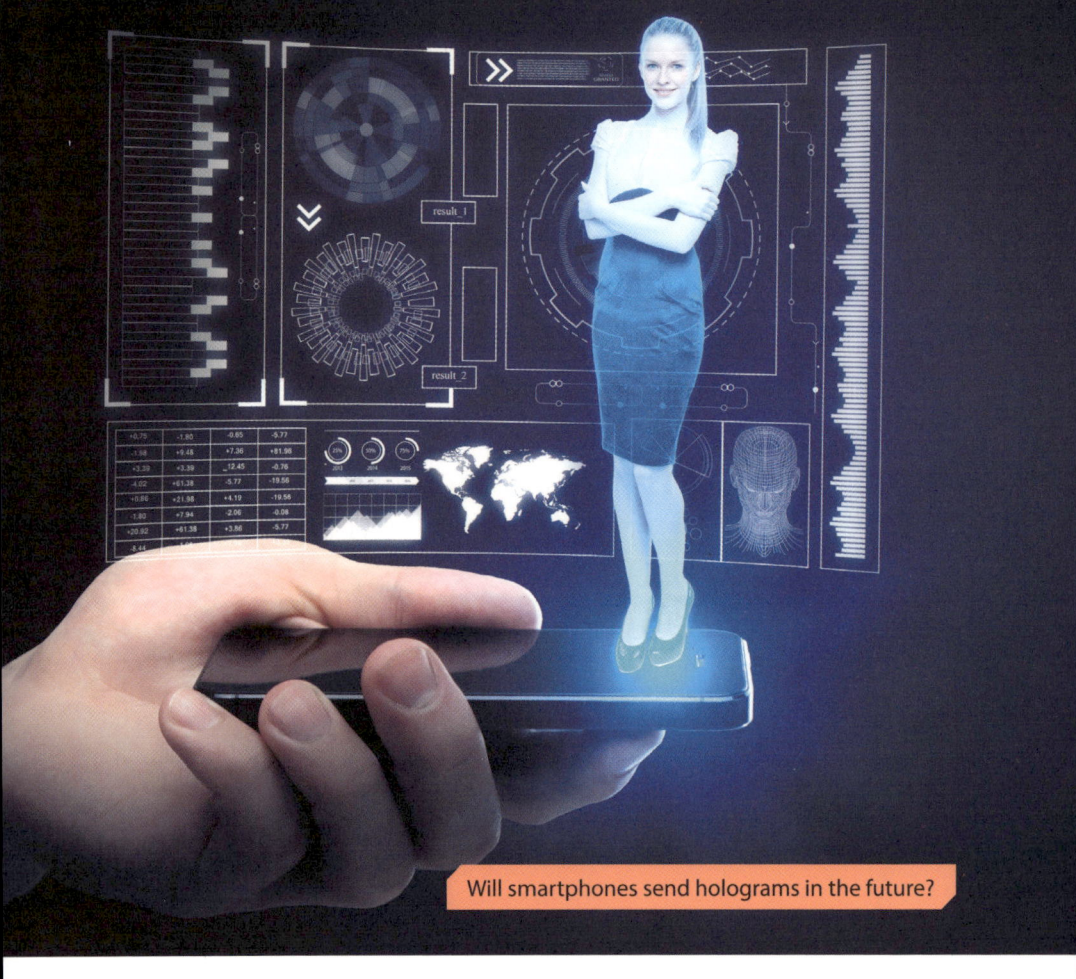

Will smartphones send holograms in the future?

Dr. Venter also made a machine that sends DNA to another machine to be "made again."

In the future, what else will doctors be able to send through computers and the internet?

A medical scientist using a microscope and computer together

Computers will continue to change medical science. Medical scientists will continue to use computers and machines to learn about the body and how it works.

A team of medical scientists

Medical scientists work in many places like universities and hospitals.

You can also find them at medical companies.

Now you know what a medical scientist is and what they do. Is it a job that you want to do? Math and science skills are important, but medical scientists must also:

☐ like to make things
☐ like to find answers
☐ be good with their hands
☐ want to help people

Working in a lab

Comprehension Questions

1. What do medical scientists learn about?
   (a) Medicines
   (b) Cells and bacteria
   (c) The body and how it works.
   (d) All of the above

2. What is something medical scientists do NOT study?
   (a) DNA
   (b) Diseases
   (c) Laboratories
   (d) Chemistry

3. What is a laboratory?
   (a) A kind of medicine
   (b) A room where scientists work
   (c) A job that medical scientists do
   (d) A kind of experiment

4. What does a microscope do?
   (a) Make small things look bigger
   (b) Keep your eyes safe
   (c) Grow new skin from cells
   (d) Send DNA to another machine

5. Our bodies are made of many . . .
   (a) arms and legs.
   (b) test tubes.
   (c) cells.
   (d) bacteria.

Glossary

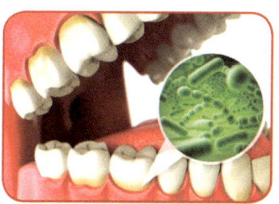

- **bacteria** (n.) small living things that can make you sick

- **biology** (n.) the study of things that are alive, like plants and animals

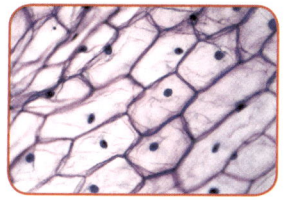

- **cell** (n.) the smallest part of a person, animal, or plant

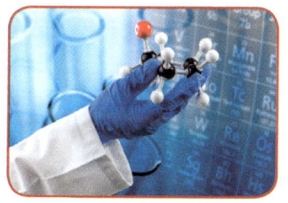

- **chemistry** (n.) the study of all substances

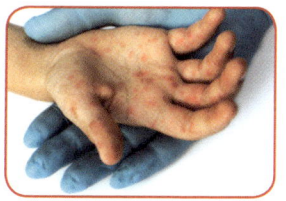

- **disease** (n.) an illness; something that makes you sick

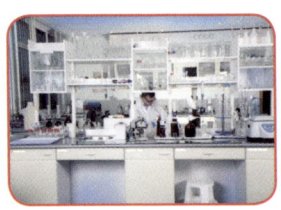

- **laboratory** (n.) a room where people do experiments

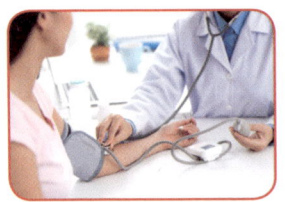

- **medical** (adj.) involving hospitals, doctors, and medicine

- **medicine** (n.) a pill or liquid taken to make a person feel better when they are sick

- **scientist** (n.) a person who studies science and does experiments

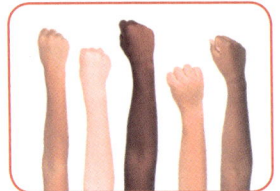

- **skin** (n.) the outer part of your body

Notes

Here are some activities, ideas, and people related to medical science. Readers may enjoy researching these ideas to learn more about the medical science field.

Growing bacteria at home: Using common items found in the kitchen, you can grow your own bacteria. What does it look like after a few days? After a week?

Images of bacteria: Bacteria comes in all different colors, shapes, and sizes. Do a search for images of bacteria and find out what parts of our homes have the most!

Craig Venter: This medical scientist in America has successfully made human cells. He is studying ways to help people live for a long time.

Lynn Margulis: This microbiologist was the first to really understand how cells develop. At first people did not believe her, but now her ideas are followed by all scientists.

Henrietta Lacks: Cells from Henrietta Lacks were used by scientists to study cancer and other diseases all over the world. There is a book and movie about Lacks and how her cells have been used in medical research.

List of Books

LEVEL 1

1. Robotics Engineers
2. Cyber Security Experts
3. Medical Scientists
4. Social Media Managers
5. Asset Managers

LEVEL 2

1. Drone Pilots
2. App Developers
3. Wearable Technology Creators
4. Computer Intelligence Engineers
5. Digital Modelers

LEVEL 3

1. IoT Marketing Specialists
2. Space Pilots
3. Water Harvesters
4. Genetic Counselors
5. Data Miners

LEVEL 4

1. Database Administrators
2. Nanotechnology Research Scientists
3. Quantum Computer Scientists
4. Agricultural Engineers
5. Intellectual Property Lawyers

"The future of the economy is in STEM. That's where the jobs of tomorrow will be."

James Brown (Executive Director of the STEM Education Coalition in Washington, D.C.)

Data from the US Bureau of Labor Statistics (BLS) support that assertion. Employment in occupations related to STEM—science, technology, engineering, and mathematics—is projected to grow to more than 9 million by 2022 [in the US alone] . . . Overall, STEM occupations are projected to grow faster than the average for all occupations.

from *STEM 101: Intro to Tomorrow's Jobs* **US Bureau of Labor Statistics**